THE CUTTING EDGE

ELECTRONICS

MP3s, TVs, and DVDs

Chris Oxlade

Heinemann
LIBRARY

H **www.heinemann.co.uk/library**
Visit our website to find out more information about **Heinemann Library** books.

To order:
☎ Phone 44 (0) 1865 888066
🖹 Send a fax to 44 (0) 1865 314091
🖥 Visit the Heinemann Bookshop at **www.heinemann.co.uk/library** to browse our catalogue and order online.

First published in Great Britain by Heinemann Library, Halley Court, Jordan Hill, Oxford OX2 8EJ, part of Harcourt Education.
Heinemann is a registered trademark of Harcourt Education Ltd.

Editorial: Sarah Shannon and Kate Bellamy
Design: Richard Parker and Tinstar Design www.tinstar.co.uk
Illustrations: Jeff Edwards & Richard Parker
Picture Research: Natalie Gray and Bea Ray
Production: Chloe Bloom

Originated Digital Imaging
Printed and bound in China by South China Printing Company

ISBN 0 431 13262 3 (hardback)
10 09 08 07 06
10 9 8 7 6 5 4 3 2 1

British Library Cataloguing in Publication Data
Chris Oxlade
Electronics (The Cutting Edge)
621.3'81
A full catalogue record for this book is available from the British Library.

Acknowledgements
The publishers would like to thank the following for permission to reproduce photographs:
Alamy pp. **9, 11, 22, 32, 38, 45, 44**; Alamy (Steven May) p. **24**; Archos pp. **4, 5, 47, 48**; Audiovox pp. **5, 20**; Corbis pp. **6, 16**; DMD pp. **4, 41**; Konica Minolta pp. **4, 31**; Napster Uk Homepage p. **26**; PalmOne p. **13**; Playstation pp. **5, 52**; Samsung pp. **5, 10, 14, 36/37**; Science & Society Picture Library p. **50**; Science Picture Library pp. **18, 34**; Sony pp. **4, 29**;

Cover photograph of NEC cell phone. The phone measures of 85 mm in width, 54mm in height and a depth of only 8.6 mm. It is said to be the world's smallest and slimmest cell phone. The photograph is reproduced with permission of Yuriko Nakao/Reuters/Corbis.

Our thanks to Ian Graham for his assistance in the preparation of this book.

Every effort has been made to contact copyright holders of any material reproduced in this book. Any omissions will be rectified in subsequent printings if notice is given to the publishers.

The paper used to print this book comes from sustainable resources.

Contents

Any words appearing in the text in bold,
like this, are explained in the Glossary.

Electronics at the edge

Electronic devices and gadgets such as mobile telephones, televisions, and **digital** cameras are part of our modern way of life. We use these "consumer electronics" every day for communication and entertainment without a second thought. Imagine for a moment your life without your CD player, your television, your mobile telephone, or your digital camera, and you will realize how much you need them.

This book is about the consumer-electronic technology that is state of the art, or at the "cutting edge", at the beginning of the 21st century. It looks at cutting-edge devices that we use at the moment and how they work. It looks at what devices came before them, and the major breakthroughs that made them possible. Finally, it looks at what devices and gadgets are likely to replace them in the near and distant future.

Speed of change

Just a few years ago, devices that we take for granted today, such as digital music players and cameras, did not exist, or they were too expensive for most people to buy. This shows just how incredibly quickly electronics are changing. This happens because people always want the latest, fastest, smallest, most powerful, most feature-filled gadgets, and technology companies continue to produce them. When a device appears in the shops, companies are already designing a new, improved version to take its place a few months later. This makes it hard to predict the future of electronics. In fact, the only thing we can be sure of is that things will change very quickly!

Electronic history

The first electronic **component** was invented at the start of the 20th century. It was a **valve** that allowed one electric current to control another. All electronic devices, such as radios and televisions, used to use bulky valves until the late 1940s. Now, almost every modern device contains one or more **microchips**. These were developed in the 1950s. Since then, manufacturers have found ways to fit more and more components onto microchips. This allows very complex electronic circuits to be fitted into a tiny space.

The large microchip in the centre of this picture contains thousands of microscopically small components.

✂ Make the connection

What are electronics?

An electronic device works using electricity, but it is different to an electric device. For example, a torch is an electric device. When you turn it on, electricity flows from the battery through the bulb. When you turn it off, the electricity stops. Other examples of electric devices are toasters, hair-dryers, and electric heaters.

Electronic devices work using electric **signals** that represent something else, such as sound for example. The devices can change, or process, the signal. A radio is an electronic device. It takes the electric radio signal, processes it, and turns it into the sound you hear. All the devices and gadgets in this book are electronic devices.

Representing information

In all the devices in this book, electricity represents information (such as sound or images). For example, in a CD player, electricity represents sound; in a television, electricity represents moving pictures. The electricity that represents the information is called an electric signal.

There are two types of electric signal – **analogue** and digital. In an analogue signal, the strength of the electricity varies. If you take an analogue signal that represents a sound wave, the changing strength of the electricity represents the changing shape of the sound wave (see diagram).

In a digital signal, the electricity is either on or off. A sequence of ons and offs represents the ones and zeros of **binary numbers**. In a digital signal representing a sound wave, these numbers represent the differing strength of the sound wave.

The numbers are recognized and processed by digital devices. For example, when a digital sound signal is processed, the numbers it contains are turned into sound.

Most devices in this book contain both analogue and digital circuits.

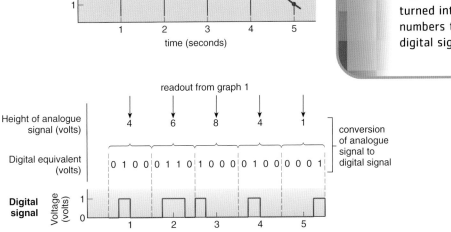

An analogue signal is measured thousands of times a second and the measurements are turned into binary numbers to form a digital signal.

Going digital

During the last two decades there has been a switch from analogue electronics to digital electronics. This change is sometimes known as the "digital revolution". But why has it happened? What are the advantages of digital electronics over analogue electronics?

The main advantage is that once information is in digital form, it can be stored as files in memory, on **hard-disk drives** and on **CDs** and **DVDs**. It can then be transmitted over digital **networks** and processed by computers. Sounds, photographs, and video clips can all be handled in the same way and on the same machine. Before digital electronics, you needed lots of different machines, as sounds were stored on audiotape, photographs were stored on film, and video was stored on videotape.

Digital circuits also mean that data can be reduced to a smaller size. This is called **compression**. It allows information to take up less storage space and to be sent from place to place more quickly. The information must be returned to its original size afterwards. For example, **MP3** and **WMA** (see page 25) are two popular **formats** for storing sounds. **JPEG** is a format for storing photographs.

✕ Make the connection

"**Bandwidth**" is a common word in the world of communications. It means the speed that information can be sent along a communications link such as a cable or an **optical fibre**. You can think of a communications link carrying information as being like a pipe carrying water. A wider pipe can carry more water, just as a link with greater bandwidth can carry information faster. Also, many more digital signals than analogue signals can fit down a link. The word "**broadband**" describes a communications system that can carry a very high number of digital signals.

Convergence

"Convergence" is one of the buzzwords in electronics. It means that one or two state-of-the-art devices can do the jobs that were done by many different devices in the past. For example, in the 1980s you needed a camera to take photographs, a cassette player to listen to music, and a telephone for making telephone calls. Now, a cutting-edge mobile telephone can do all of these jobs. This is possible because all the different types of information are handled in digital form.

In the middle of the 1990s nearly all photographs were recorded on rolls of plastic film (left). Today most photographs are recorded digitally on memory cards such as Compact Flash (below).

Going mobile

Mobile telephones are one of the fastest-growing electronic gadgets in history. In the mid-1990s, a mobile telephone was rarely seen. Now, in many countries, more than three-quarters of the population own a mobile. In these countries, there are now as many mobile telephones as there are fixed-line telephones in homes and offices. The mobile telephone is a success because it is really handy for people on the move, it is fun to keep in touch with friends, it is really helpful in an emergency, and it is a great handheld games machine.

Mobile workings

A mobile telephone is simply a telephone combined with a radio **transmitter** and **receiver**. Voice calls and other information are carried to and from the telephone by radio signals. The signals travel between the telephone and a mast, tower, or building, called a base station.

Most mobile phones have a similar arrangement of buttons for dialling, navigating menus and inputting text.

Landmarks in time

1876 Alexander Graham Bell demonstrates the first working telephone

1878 The first telephone exchange, with 21 lines, is opened in New Haven, Connecticut, USA

1920s Police in the USA begin using radio telephones

1946 The first car-based radio telephone network opens in the USA

✕ Make the connection

There have been three important steps in the development of today's mobile telephones. The first step was the development of radio telephones in the 1940s. These were so big and needed so much power that they could only be used in cars. The second step was the invention of a handheld telephone that used a cell system. This was invented by Dr Martin Cooper of Motorola. He made the very first call on a portable mobile telephone in 1973. Early mobile telephones were as heavy as house bricks, and nearly as big! The final step was the change from analogue to digital, which happened in the early 1990s. This allowed a far greater number of telephones to use the system.

Each base station handles signals for mobile telephones in the area around it. This area is called a cell, which is why mobile telephones are also called cellphones. In city centres, where there are many mobile telephones in a small area, cells are smaller than in the countryside.

Base stations are connected together to form a mobile network. This is connected to the general **telecommunications** system so the phone can make calls to other mobile networks, fixed-line telephones, and the internet.

Mobile telephone towers at a base station.

1973 The first portable mobile telephone is demonstrated by Motorola

1978 The first cellular telephone networks begin operation

2003 The first third-generation (3G) mobile phones become available

Where are we now?

The state-of-the-art mobile telephone is a remarkable feat of electronics. It can send and receive text messages, take and display colour photographs and videos, play music, and receive radio stations. And it also works as a telephone!

Getting the message

Mobiles use two types of message service: text messaging and Multimedia messaging. Text messaging or SMS (short message service) allows short messages to be sent.

Nobody predicted that texting would become so popular, or that it would give rise to a new text language. Multimedia messaging (MMS) is more advanced. It allows photographs and video to be sent from mobile to mobile.

Photographs and video are taken by a built-in camera that is as good as a cheap digital camera. Pictures are stored in the memory so they can be sent later. The video camera also allows two people to see each other as they are talking. This is called video conferencing.

✕ Make the connection

Mobile dangers

A long-running debate about the danger of using mobile telephones continues. Mobiles use high-frequency radio waves. These are similar to the waves used in microwave ovens, but are much weaker. We know that some types of radio waves can damage the cells in our bodies. Some scientists believe that there must be a danger, especially for children, of brain damage from a radio transmitter held right next to the head. Despite several research projects, no proof of damage has been found. However, mobile-telephone users who wish to be cautious limit the number of calls they make, use a hands-free set, and text instead of making voice calls when possible.

Mobile fun and games

The latest mobile telephones are also mini entertainment systems. Most have built-in games and can also **download** games from the internet. Some play MP3 music files (see page 25 for more about MP3), and others are also radios. Most basic mobiles can also browse the internet – although they can only open simple, text-based web pages using WAP (wireless access protocol).

Smartphones

A **smartphone** is a mobile telephone combined with a PDA (portable digital assistant), which is a type of **palm-top computer**. As well as being a mobile telephone,

a smartphone allows a user to surf the web, send and receive emails, and use applications such as a diary and word-processor on the move. It is an example of converging technologies of the worlds of the mobile telephone and the personal computer.

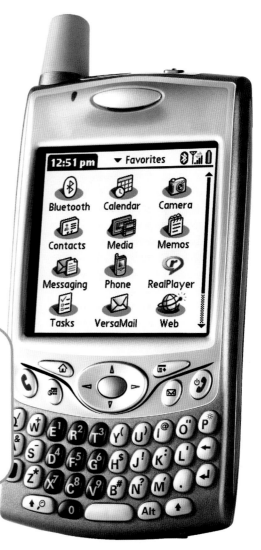

The palmOne Treo 650 is a cutting-edge smartphone. It allows users to send and receive emails, and view attachments on the high-resolution screen. It is also a digital camera and an MP3 player.

Inside-out mobile

Let's see what interesting electronics can be found in a mobile telephone.

Antenna: This works as a transmitter to send radio signals to base stations. It also work as a receiver to collec radio signals from a base station.

LCD display: The liquid-crystal display **(LCD)** shows text and **graphics**. Some phones have two displays, one especially for photographs and video.

Speaker: This turns electrical signals into sound at the earpiece.

Keyboard: This is used for entering phone numbers, text, and other data, and for operating the phone. It sends information to the processor.

There are also lots of electronics hidden inside the mobile phone case:

Battery: A rechargeable battery provides power to the phone's circuits. Most of the power is needed to send radio signals from the transmitter.

Amplifiers: These strengthen signals before they go to the antenna and when they arrive from the antenna.

Microprocessor: This controls all the functions of the phone by carrying out programmed instructions. It is capable of millions of calculations a second.

Digital-signal processor: This codes and decodes signals to and from the antenna.

Memory: This contains instructions for the microprocessor. It is used for storing address-book data, text, photographs, and video.

SIM card: The SIM card stores information the phone needs to connect to a particular mobile network.

Second and third generations

Every few years the technology of mobile telephones changes radically. The new telephones are said to belong to a new generation. The current generation is the third-generation. These are known as 3G phones, but there are still millions of second-generation (2G) phones in use.

Each new generation of phones can send and receive more data than the last. This is important for **applications** such as photo and video messaging and web browsing. These require a great deal of data to be sent quickly. The 2G phones are too slow for these applications, but the 3G phones are much faster.

Each new generation of mobile telephones communicates with base stations in a different way from previous generations. So a whole new network is needed.

>> What is the future?

Mobile television

Soon some state-of-the-art mobile telephones will also work as handheld televisions. They use a signal format called Digital Video Broadcasting – Handheld (DVB-H). The television signals do not come over the telephone network. They come direct from television transmitters. The pictures are much clearer than the current small-screen LCD televisions.

Exciting developments

What will the cutting-edge mobile telephone look like in ten years time? One prediction is that the mobile telephone will no longer exist. Instead, it will be a super smartphone – a machine that will do the jobs of all the different gadgets we carry today.

PACE

Some experts use the word PACE for such a machine. PACE stands for "personal assistant, communications, and entertainment". It would be a small computer, a mobile phone, a camera, a music and video player, a radio, a television, and a storage device for photographs and other personal data.

Mobile teeth

Mobile telephones will continue to get smaller. There are already devices that you can wear like a wristwatch. In the future they could be embedded in spectacles, or even in your teeth! Other future phones will need large screens to display photographs and games. Some screens will be made of electronic paper, which can be folded and rolled up. LED screens already exist. These are thinner, lighter, and clearer than standard mobile-phone screens.

The future network

A completely new network will be needed to deliver all the services to the next generation of mobile telephones. This will be the fourth-generation (4G) network. It will begin operating some time around 2010. It will work in a very different way to the current network. Instead of using cells and base stations, it will work like the internet does now. Telephones will be permanently connected to the network. Calls, text messages, and so on will move as packets of information, with each telephone passing the packets on. Speed will be "ultra-broadband" – many times faster than even today's broadband internet services.

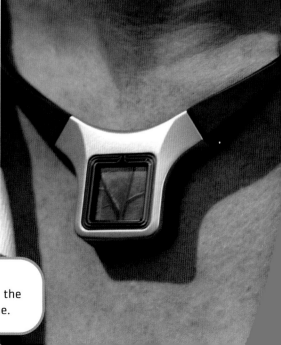

One idea for how mobile telephones might look in the future - a jewellery phone.

The new phones will no longer have a number, but an internet-style address. Future telephones will be able to connect to the communications network through the mobile network when outdoors, through **Wi-Fi** networks at home and at "hot-spots" in hotels and airports. They will also be able to connect to the normal telephone network at home through a **Bluetooth** wireless link.

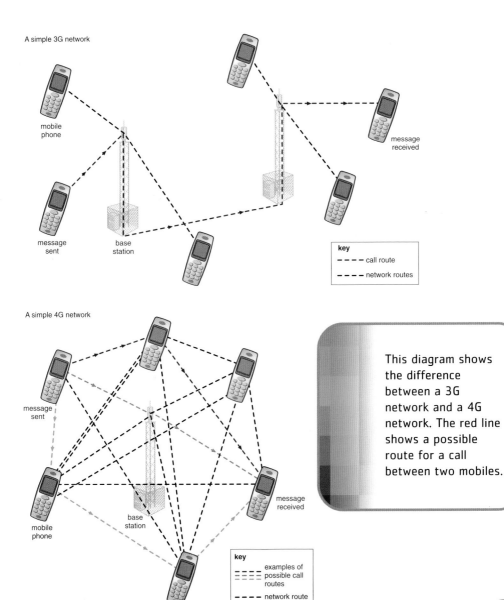

A simple 3G network

mobile phone

message sent

base station

message received

key
- - - - call route
- - - - network routes

A simple 4G network

message sent

mobile phone

base station

message received

key
- - - - examples of possible call routes
- - - - network route

This diagram shows the difference between a 3G network and a 4G network. The red line shows a possible route for a call between two mobiles.

Look, no wires!

When we hear the word "radio", we normally think of radio **broadcasting**, which brings us music, news, sport, talk, and other entertainment. "Radio" is also used for the devices that we use to listen to radio broadcasts. These are properly called radio receivers.

Making waves

Radio uses radio waves. These are invisible and travel through the air at the speed of light. The radio waves are created by a transmitter. They are sent out into the air and are picked up by an antenna in a radio receiver.

Radio is also used in two-way radios, such as walkie-talkies, and aircraft and ship radios. In these systems, the radio waves travel directly from one handset to another. Each handset has a transmitter and a receiver.

Radio waves also provide vital links in many communications networks. For example, radio waves carry signals to and from mobile phones.

A modern analogue radio receiver. It detects the radio waves for the required station and turns the signal back into sound.

Landmarks in time

1864 James Clerk Maxwell predicts the existence of radio waves

1888 Heinrich Hertz proves that there are radio waves

1896 Guglielmo Marconi sends a Morse code message between two aerials by radio

1906 Reginald Fessenden works out how to send sounds and voices by radio signals

Make the connection

Marconi's marvel

Radio waves were discovered in the late 19th century. Many scientists investigated them, but one man, an Italian called Guglielmo Marconi (1874–1937), pioneered their use in communications. Marconi began experimenting in 1894 in Italy, and then moved to Britain. He transmitted a simple signal in 1896. In 1901, he demonstrated the potential of radio by sending a Morse code signal across the Atlantic. Marconi was awarded the Nobel Prize for Physics in 1909.

The radio spectrum

There is a whole range of radio waves. This is called the radio spectrum. The waves have different frequencies. Mobile telephones, broadcast radio and television, and wireless computer networks use waves of different frequencies. This ensures that they do not interfere with each other.

Different radio stations broadcast on waves with different frequencies. Receivers pick out or "tune in" to the required station. Today, there are so many frequencies being used that the whole radio spectrum is used up. New services have to wait until old ones are switched off, freeing up the waves.

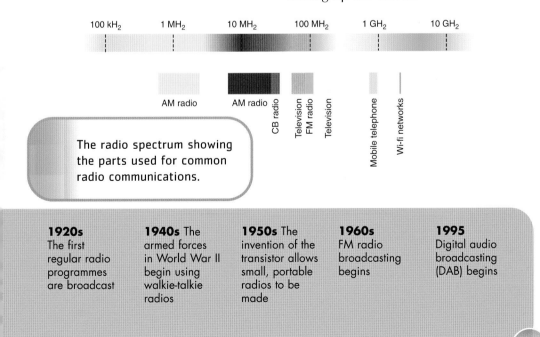

The radio spectrum showing the parts used for common radio communications.

1920s	1940s The	1950s The	1960s	1995
The first regular radio programmes are broadcast	armed forces in World War II begin using walkie-talkie radios	invention of the transistor allows small, portable radios to be made	FM radio broadcasting begins	Digital audio broadcasting (DAB) begins

Where are we now?

In the early 21st century, the world of radio broadcasting is split between analogue and digital. Most broadcasts are still made using analogue radio signals, but digital signals are now being broadcast at the same time.

Analogue radio

In analogue radio broadcasting, signals are broadcast in two different ways. These are called AM (which stands for **amplitude** modulation) and FM (which stands for frequency modulation). Most radio receivers can detect both AM and FM broadcasts. FM can be broadcast in stereo. This means that two different channels of sound are broadcast at the same time. This produces a better sound. However, both analogue signals are prone to interference. This is heard as hiss and crackle.

In some countries, simple programme information is also sent along with the sound. This allows the radio to show the name of the radio station it is receiving. The system also allows traffic announcements to interrupt the station that is already playing.

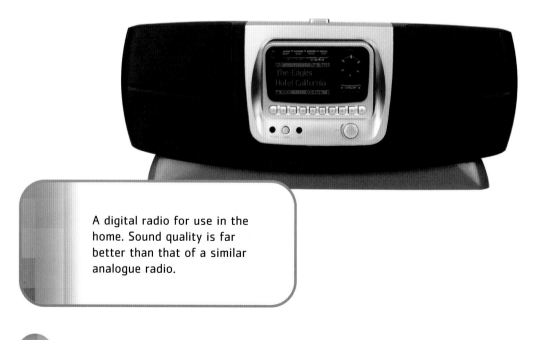

A digital radio for use in the home. Sound quality is far better than that of a similar analogue radio.

>> What is the future?

Digital radio will eventually take over from analogue. When most people own digital radios, the analogue signals will be switched off. This will take many years because of the huge number of radio receivers in the world. In the UK, it is expected that half of the radio receivers will be digital by 2010.

In the future, digital radios will be incorporated into other devices, such as mobile telephones and MP3 players. Digital radios combined with mobile telephone technology will mean that radio can be interactive. Listeners will be able to vote for songs, answer quizzes, and so on directly via their radio.

Digital radio

In digital radio, the radio signals that are broadcast are digital signals. The signals are much simpler than in analogue radio, so they are less prone to interference. The signals are also reduced in size (compressed) using a code. This cuts the amount of data that needs to be sent by up to ten times. Digital radios usually give clear sound with no hiss and crackle. Digital radio stations also broadcast data such as artist and track names. These appear on the radio's display.

Digital radio signals can be broadcast via satellite, cable, and internet. Some broadcasters allow radio programmes to be downloaded as MP3 files. Sometimes they even upload them automatically to a listener's PC. This is known as "podcasting".

Music

The path to iPod

The iPod, made by the Apple computer company, is a digital music player at the cutting edge of music technology. But how do they work, how were they developed, what are the alternatives, what came before, and what will the future bring?

Record and play back

We cannot capture sound waves themselves. We have to represent them in some way to record them.

First, a microphone turns the sounds into electrical signals. The signals are then recorded. For example, on a cassette tape, the signal is recorded as a magnetic pattern on the tape. The pattern represents the pattern of the sound. A tape player detects the pattern and rebuilds the signal from it. Most music is recorded in stereo, so there are two channels of music in a recording. These are played at the same time.

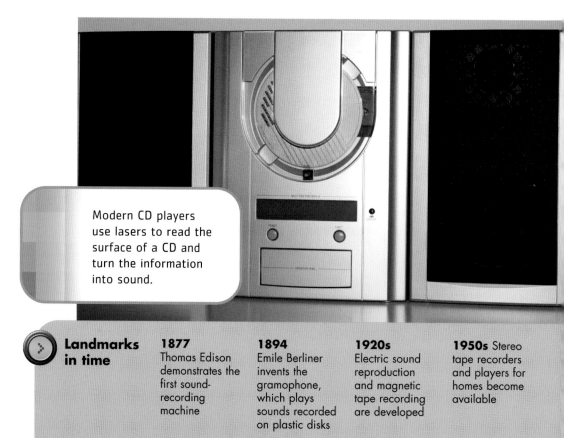

Modern CD players use lasers to read the surface of a CD and turn the information into sound.

Landmarks in time

1877	1894	1920s	1950s
Thomas Edison demonstrates the first sound-recording machine	Emile Berliner invents the gramophone, which plays sounds recorded on plastic disks	Electric sound reproduction and magnetic tape recording are developed	Stereo tape recorders and players for homes become available

✂ Make the connection

Wax to CD to MP3

The first sound recordings were made mechanically. This was before electronic devices had been invented. In the 1870s, the first sound recorder, the phonograph, recorded sounds by cutting a wavy groove into tin foil with a needle.

Electronics were first used to record sound in the 1920s with the invention of the microphone and magnetic tape. Tape remained the main way of recording until digital and optical recording was developed by Philips and Sony in 1982 in the form of the compact disc (CD). The CD had taken over from tapes and vinyl records by the middle of the 1990s.

The MP3 format was first developed in the early 1990s. It did not become popular, however, until music tracks began to be published in MP3 format in 1999.

Most sound recordings are now digital. Instead of recording the sounds as a pattern of electrical signals, the signal is made into a line of zeros and ones to represent the sounds. These are recorded on a CD, on a hard-disk drive, and sometimes on tape. A player, such as an MP3 player, reads the numbers and turns the digital signal back into sound.

1978 Sony releases the Walkman, a portable cassette player

1982 Compact disc (CD) digital audio is launched

1988 The MP3 digital sound format is developed

1999 The first portable MP3 players become available

2004 The music industry launches download charts

Where are we now?

Today, there are more ways to record and play music than ever before. The personal computer has made it possible for anybody to make their own recordings. There are many different ways to make a digital recording. It is also still possible to record using analogue tapes and records, and some people prefer listening to their music this way.

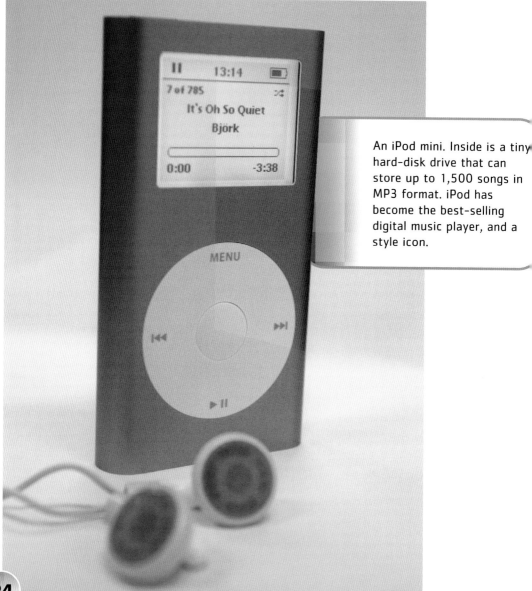

An iPod mini. Inside is a tiny hard-disk drive that can store up to 1,500 songs in MP3 format. iPod has become the best-selling digital music player, and a style icon.

Music players

Music players range from giant hi-fis to tiny digital players that are not much bigger than a finger. Larger machines play music through speakers. Smaller, portable players play through earphones.

Older CD players can play only pre-recorded CDs in the original CD-DA format. State-of-the-art CD players can also play files recorded onto **CD-Rs** and **CD-RW** discs in different file formats, such as CD-DA, MP3, **WMA**. Having a CD player that can play these different file formats means that a CD can have several different types of music album recorded on to it.

MP3 players play MP3 files, and often WMA too. The files are stored in the memory or on tiny hard-disk drives. Up to 10,000 songs can be recorded. The files are transferred to and from the players through a cable. The latest MP3 players can record MP3 files straight from CD players, without using a PC.

Sony's MiniDiscs (MDs) are small discs, like mini CDs. The latest version, called Hi-MD, can store about 500 songs in ATRAC format. They can also play other digital music files, such as MP3.

Digital recording methods

Compact disc digital audio (CD-DA) is used on compact discs. A typical three-minute music track takes up about 30 **megabytes** of storage space on a CD. This is the original format of digital recording.

An MP3 file is a compressed file (a file reduced in size). An MP3 file for a typical music track takes up about 3 megabytes – ten times less than a CD-DA file. MP3 is short for MPEG audio layer 3. An alternative to MP3 is Windows Media Audio (WMA). This works in the same way but has been developed by Microsoft. MP3 and WMA files can be saved and copied like any other computer file.

ATRAC has been developed by Sony for use with its MiniDisc system. It compresses a typical music track to about one megabyte. The quality of the sound is still almost as good as on a CD.

Downloading and ripping

So where do MP3 and other digital music files come from? They can be either copied from the internet or taken from CDs or other computer files. Copying from a computer on the internet is called downloading. Most digital music users download the files to their PCs. Files recorded from CDs and converted to MP3 or WMA files are also saved on a PC. This is known as "ripping". Files can then by played on the computer, downloaded to a digital music player, or "burned" onto a CD.

Downloading music files has become big business. The number of downloads from online music stores such as the Napster and iTunes Music Store went up ten times in 2004 alone. The sales of music are now measured by the number of downloads as well as the number of CDs sold. The "download chart" is as important as the CD sales chart.

A screen grab of the Napster UK hompage. Napster is one of the most popular sites for music downloads.

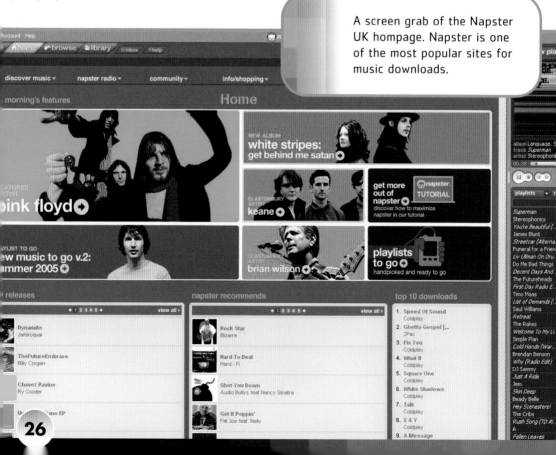

Smaller and better

Now that MP3 and WMA have become popular, music players are likely to change at a steady pace rather than dramatically. Music players will continue to grow in capacity, so that they will hold more songs, but they will get smaller. Hard drives that are smaller and more efficient will save energy, and so batteries will last longer. Players may be sold already loaded with music or with talking books. One day, they may even be cheap enough to be disposable.

Pirating music

Illegal copying of digital music (called pirating) is a big issue. Most of the music that you listen to is copyrighted. That means that each copy of it must be paid for. The money goes to the record company and the artist or band. If you buy a CD, you are allowed to copy files from it to play on your own MP3 player, but you are not allowed to give them to somebody else. In the same way, you are not allowed to copy MP3 files from a friend's computer, or from internet sites, without paying for them. It is illegal. Music companies argue that it is the same as stealing a CD from a shop. In future, digital music files that you buy online may self-destruct after a certain number of plays!

>> What is the future?

In the future, music players will also be able to play DVDs, video, and other new formats. These new machines will become personal media players (**PMPs**) and will provide a whole range of entertainment. Digital music players will be incorporated more and more into mobile phones, cameras, **memory sticks**, and so on. Players will also be able to receive digital radio broadcasts. The next digital music format is likely to be Advanced Audio Coding (AAC), nicknamed MP4.

Megapixel magic

Cutting-edge digital cameras are incredible machines full of amazing electronics. A modern camera still does the same job that cameras have done for nearly 200 years – that is to record an image of the scene it is pointed at. It is still up to the photographer to take interesting photographs!

Digital cameras have improved in quality so much that they are no longer just used for taking party and holiday snaps. Professional photographers use digital cameras to take photographs for newspapers, magazines, websites, adverts, books, and catalogues. They normally use sophisticated cameras with a wide range of features to take high-quality photographs quickly.

How cameras work

All cameras have a lens at the front. The lens collects light coming from a scene, and focuses it to make a small image of the scene in the back of the camera. In a film camera the image is recorded on light-sensitive film to make a photograph. In a digital camera the image is recorded electronically by a device called a **CCD** (see page 31). The image is stored in the memory. A photograph can then be reproduced from the camera's memory.

Most cameras also have a shutter and an aperture. The shutter opens to let light in through the lens. The aperture is a hole behind the lens that changes in size to control the amount of light that gets through the lens.

✂ Make the connection

Digital cameras are also used very widely in science, industry, medicine, and other fields. For example, astronomical telescopes contain digital cameras. These record images of the stars, satellites in space, weather patterns, sea temperatures, and other information. Space probes take images of other planets with digital cameras. In many countries, digital cameras are used to photograph cars that exceed the legal speed limit.

In an electronic "point-and-shoot" camera, the picture-taking process is controlled automatically. When you press the button to take a photograph, the camera controls the whole operation. The lens moves in or out to focus the image, making it crisp and clear. Then the camera uses an electronic eye to measure how bright the scene is (how much light is coming into the camera). The camera adjusts the shutter speed and aperture size so the right amount of light will reach the CCD or film. This makes sure the photograph is neither too light nor too dark. The system also activates the flash if the scene is too dark. Finally, it opens the shutter to take the photograph.

A state-of-the-art digital camera. It features a high-quality zoom lens, a flash, and is fully automatic.

The history of camera technology

The first cameras had no way of recording images at all. They made an image on a screen and were used by artists for copying scenes. Methods of recording images with light-sensitive chemicals were invented in the 1820s, and photography was born. Nearly all cameras had manual controls until the 1960s. Then electronic controls were introduced. Autofocusing (where the camera focuses the lens) was invented in the 1980s. This allowed the first point-and-shoot cameras to be made.

The biggest change in camera technology is currently underway. Although electronic image recording was developed in the 1950s, it was very expensive until the 1990s, when digital cameras became available. Since then, prices have fallen quickly, and the detail that the camera can record (called the resolution) has increased dramatically. The detail in photographs taken by professional digital cameras is now the same as the detail available from traditional film cameras.

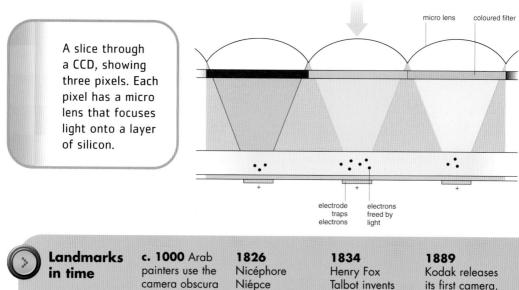

light from lens

micro lens   coloured filter

A slice through a CCD, showing three pixels. Each pixel has a micro lens that focuses light onto a layer of silicon.

electrode traps electrons

electrons freed by light

Landmarks in time

c. 1000 Arab painters use the camera obscura as an aid for painting scenes

1826 Nicéphore Niépce takes the first photograph using paper sensitive to light

1834 Henry Fox Talbot invents the negative/positive system for taking and printing photographs

1889 Kodak releases its first camera, ready loaded with a roll of plastic film

CCD imaging

Digital cameras record images using a charge-coupled device (CCD). This is a special type of microchip. It acts like the film in a film camera to record the photograph. One side of a CCD is covered with a grid of tiny dots called **pixels**. The more pixels on the CCD, the greater the resolution it has, so the more detail it can record in the photograph.

Each pixel contains thousands of silicon atoms. When light hits the atoms, they become electrically charged. The brighter the light, the more charged they become. To record the image, electronic circuits measure the charge in each pixel and record the reading in the camera's memory. CCDs are more sensitive to light than traditional film, so digital cameras can take photographs in darker conditions.

Where are we now?

Digital cameras are taking over from film cameras at an incredible pace. Digital cameras have the advantage that taking a photograph is free (you only have to pay for processing and printing). You can take as many photographs as you like and just keep the good ones. It is estimated that 250 million digital images are taken around the world every day.

A state-of-the-art compact digital camera has a CCD with four or five **megapixels**. It can take superb, sharp, bright photographs fully automatically. It can also record short movie clips. Photographs are displayed on a bright LCD display.

A state-of-the-art professional digital camera has an even larger CCD, perhaps with eight or more megapixels. It allows the photographer to have complete control over the aperture, shutter, and flash settings, but it can work fully automatically. It also accepts different lenses and accessories such as flashguns.

This professional digital camera gives the photographer much greater control than a point-and-shoot digital camera.

1930s	1978 Konica	1990 The	1994 The JPEG	2004 Digital
Electronic light measuring is introduced into cameras	launches the first electronic point-and-shoot camera with automatic focusing and light metering	Kodak DSC 100 is the first digital stills camera to come onto the market	format is developed for storing compressed digital images	cameras are so popular that Kodak stop making film cameras

Storing images

Images in a digital camera are stored as files in the memory. These are normally in JPEG format. The JPEG format reduces the size of the files so that lots of files can be stored in the memory.

Some cameras have built-in memory, but most have a removable memory card, such as a Compact Flash card or Multimedia card. A few cameras can also store images on recordable CDs. Cameras can only store a certain number of images before the memory becomes full. So images are transferred to another device, usually a computer, for storage.

Imagine being on holiday with a digital camera. You want to take hundreds of photographs, but you do not want to carry a laptop PC with you. You need a device called a portable media player (PMP). It stores files on a tiny hard-disk drive and displays photographs on an LCD screen.

An iPod photo. This device features a 60GB hard-disk drive. It can store 15,000 MP3 songs or 25,000 photographs, which can be displayed on the high-resolution colour screen or on a television.

Files can be uploaded from a camera directly to it. Many portable media players can also be used as digital music players. Some can play digital video clips. **Uploading** and sharing photographs will become easier with cameras that connect to wireless networks. This will allow direct emailing of photographs from the camera.

Cameras everywhere

There are already low-resolution digital cameras built into devices such as mobile phones. Falling costs will soon allow them to be put into other devices, too, such as music players or **GPS** receivers. Despite this, professional photographers will still need a dedicated camera with a large, high-quality lens.

>> What is the future?

The basic technology of digital cameras is already with us. It is unlikely that there will be any dramatic advances in the near future. However, digital cameras will gradually evolve. We can definitely predict that the resolution of cameras will continue to increase. Eventually, ordinary point-and-shoot cameras will have ten megapixels or more, and prices will continue to fall. Cameras will also have hard-disk drives and large, roll-out screens to display images.

One change may come in camera electronics. There could be a switch from CCD technology to CMOS technology for recording images. CMOS stands for complementary metal-oxide semiconductor. CMOS microchips detect light in a different way to CCDs. They are cheaper to make and use less power. Other electronic components can also be built onto the same chip. This would allow cameras to have fewer parts. Perhaps it will allow for disposable digital cameras to be built one day.

Pictures on plasma

What is television?

The word "television" really means seeing a live moving picture in one place of an event that is happening somewhere else. The word was made up when the first television was invented. Today, we only see this happening when we watch something live on the television, such as a football match. However, we also use the word "television" to mean the box of electronics that displays television pictures. Its proper name is a television receiver. In many countries there is a television receiver in almost every home, and often there are two or three!

A television receiver does several jobs. It shows programmes that are broadcast by television companies, such as the news. Some of these are live, but most have been recorded earlier. A television receiver also displays pictures from other devices in our homes, such as DVD players, video recorders, and games consoles.

A large widescreen plasma television. Plasma technology allows manufacturers to build giant, flat televisions.

Landmarks in time

1884 Paul Nipkow invents a spinning disk that divides a scene into lines. It works as a simple television camera	**1923** Vladimir Zworykin develops the iconoscope, an electronic device at the heart of early electronic television cameras	**1926** John Logie Baird demonstrates a working television system using the Nipkow disk	**1929** The first television broadcasts are made in the UK

From one to many

Television broadcasting means sending television pictures from one place to many television sets at the same time. In most countries, there are several broadcasting companies. Each of these offers a choice of several different television channels.

Closed-circuit television

Closed-circuit television (CCTV) is used for security in shops, offices, and other buildings. In a CCTV system, signals from the cameras are passed directly to television screens. They are not broadcast to the public.

✕ Make the connection

How does television work?

Television pictures (and the sound that goes with them) are broadcast in the form of electrical signals. The picture is made up of thousands of tiny dots of colour. The signal contains information about the colour and brightness of each dot. Each dot is red, green, or blue. These three colours are mixed in various amounts to produce the image on the screen.

The picture signal must get from the broadcaster to the television receiver in your house. The signals can travel in different ways - as electrical signals along cables, or as radio signals from a transmitter or a satellite.

Hundreds of programmes are broadcast at the same time. A television receiver, or a set-top box attached to the television, picks out the signal from the channel you want to watch. This is called tuning. It is done using an electronic circuit called a tuner. The television receiver then turns the signal back into a moving picture with sound.

1953	**1981**	**1983**	**1990s**
Colour television broadcasting begins in the USA	The first high-definition television (HDTV) system is demonstrated	Satellite television broadcasts begin in Europe	Digital television broadcasting begins

History of television

The first working television system was developed in Britain by John Logie Baird. It broadcast the first television programmes in the 1930s, using cameras and receivers with many mechanical parts. The pictures were very fuzzy and wobbly. To begin with, all broadcasts were in black and white. In the 1950s, colour broadcasting and television receivers that could display colour pictures were introduced. The most radical change to television since then has been the introduction of digital television broadcasting via cable and satellite. This happened in the 1990s. LCD screens were also developed in the 1990s as an alternative to **cathode-ray tube** television.

Where are we now?

This is a time of great change in television. New ways of broadcasting pictures have been introduced. New technologies are being used to display pictures on television receivers.

At the moment, television channels are broadcast in three different ways. Terrestrial television is broadcast using radio waves sent out from transmitters on the ground. Satellite television is broadcast using radio waves transmitted from satellites above the Earth. Cable television, however, is broadcast as electrical signals which travel through cables under the ground.

✕ Make the connection

A major breakthrough was the development of the electronic cathode-ray tube (CRT). This was a new way of producing television pictures from the electric signal. The CRT had no moving parts, so it gave much clearer images than the previous mechanical system. It was soon used in all electronic television cameras and receivers.

Nearly all television receivers can receive terrestrial broadcasts through an aerial. An extra receiver, either built into the television or in a separate set-top box, is needed to decode cable and satellite signals. The signals are then passed along a cable to the television for display.

Television channels are also broadcast using digital signals. Analogue signals are now only used for terrestrial broadcasts, and will soon be switched off.

Digital broadcasting offers clearer pictures and interactive services. It uses less power at the transmitter. The simplicity of the signals allows many more channels to be broadcast at once.

Some broadcasts are also made in widescreen format, and others include stereo sound or **surround sound**. Extra information is added to the signal for these features. Most modern televisions can recognize all these signals.

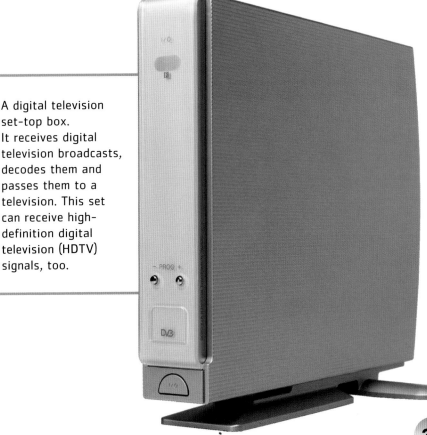

A digital television set-top box.
It receives digital television broadcasts, decodes them and passes them to a television. This set can receive high-definition digital television (HDTV) signals, too.

Screen technology

Television receivers use different technologies to display television pictures. Most televisions still use a cathode-ray tube (CRT). This is a glass tube with the television screen at its base. On the back of the screen are thousands of red, green, and blue dots. They are made to glow by firing beams of electrons at them from the end of the tube.

A close-up of an LCD television screen. You can see the columns of coloured LCD crystals. An image is made up by setting the brightness of each crystal.

The CRT now faces a challenge from flat-panel screen technologies. There are two main types of flatscreen – LCD displays and **plasma screens**. In both types of flatscreen, the screen is made up of thousands of individual dots called pixels. Electronics in the television change the brightness and colour of each pixel to show the picture. Because there is no glass tube, televisions that use these technologies can be flat. This means that they can be hung on the wall like a cinema screen. They are also lighter than CRT televisions, and they use less energy.

✕ Make the connection

LCD faults

Large LCD screens are difficult to make. There are millions of microscopically small crystals and other components. If even a tiny percentage of the pixels have a fault, the screen is rejected by the manufacturer. This happens to nearly half the screens made. It is why large LCD screens are expensive to buy.

LCD televisions

The letters LCD stand for liquid-crystal display. In an LCD television screen each pixel has a **transistor** and a liquid crystal. A liquid crystal is made of material that changes shape when electricity is applied to it. Sending a charge to one of the transistors in an LCD applies electricity to its crystal. The crystal changes shape, which makes it twist. As it twists, the amount of light that can get through the crystal changes, and this changes the brightness of the pixel. Each pixel is made up of three sub-pixels, with filters to make red, green, and blue light. The light is mixed in various amounts to produce the colour seen on the screen.

LCD televisions do have a problem. There is a tiny delay as the pixels change brightness between one frame of the picture and the next. This means that fast-moving objects in a picture often leave a faint trail behind them. This is known as image lag. It is a particular problem on very large LCD screens.

Plasma displays

Plasma displays are a state-of-the-art way of displaying television pictures. They are only 10 to 15 centimetres thick, but can be more than 1.5 metres across. They are brighter and give better colours than LCD television. Also, they do not suffer from image lag. However, they are very expensive! One minor problem with plasma screens is that if a still image is displayed for a long time, the image can become "burned" into the cells. It never disappears.

Plasma and pixels

In a plasma display, each pixel is made up of a tiny glass chamber filled with xenon or neon gas. When electricity is passed through the chamber, the gas turns to plasma. This makes the pixel glow. The brightness of each pixel depends on the amount of electricity passed through the chamber. The display is controlled by a computer. As with an LCD display, the plasma can glow blue, green, or red to produce the images seen on the screen.

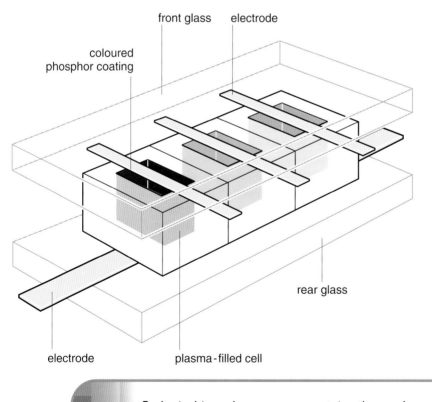

front glass   electrode

coloured
phosphor coating

rear glass

electrode   plasma-filled cell

Each pixel in a plasma screen contains three coloured sub-pixels. When electricity is applied to the electrodes, the plasma gives off particles, which causes the phosphor coatings to glow the appropriate colour.

Projection television

In a projection television, a small picture is created inside the television. This is then projected through a lens onto the screen. Most projection televisions are rear-projection televisions, where the picture is projected onto the back of a translucent screen. Projection televisions were developed in the 1970s to produce very large screens, but the pictures were quite dim and blurred at the edges. However, cutting-edge projection systems are far superior, and give bright, clear pictures.

In some projection televisions the picture is created on an LCD screen. A bright light shines through the LCD, then through the lens and the screen. State-of-the-art projection televisions are reflective instead. The light shines onto the picture, then reflects back off the picture through the lens and screen.

DLP

One reflective system is known as digital light processing (DLP). The picture is created on a microchip covered with a grid of microscopically small mirrors. This is called a digital micromirror device (DMD). Each mirror represents a pixel in the picture and is about one fifth the width of a human hair. Tiny electrical currents in the chip move the mirror from side to side. A light source is fixed on the mirrors. When the mirror moves, it either moves towards the light source or away from it. If it moves towards the light source, the mirror reflects light, making a light pixel. If the mirror moves away from the light source, the mirror does not reflect light, so it is a dark pixel. DLP is an example of a microelectromechanical system (MEM) system.

High-definition television

The next major change is likely to be a gradual switch to high-definition television. Standard television pictures are made up of 525 or 625 horizontal lines. High-definition television (HDTV) pictures have up to 1080 lines. The pictures have much more detail than standard pictures, making them incredibly clear. HDTV is broadcast in digital form, with high-quality surround sound (see page 49).

High-definition television is not a new technology. It was introduced in 1990 in Japan, and many state-of-the-art televisions can receive HDTV. However, at the moment, its use is very limited because very few television companies broadcast HDTV. Broadcasters need to invest in HDTV cameras and other systems before HDTV becomes popular. This is likely to be after 2010.

The same area of a an analogue television picture and a digital HDTV picture. There is much greater detail in the HDTV picture.

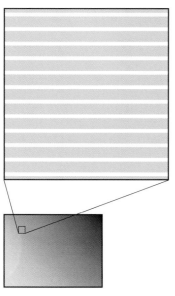

Analogue TV

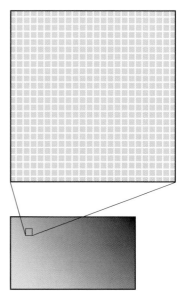

HDTV

Television on demand

In the 1950s, the only way to see a whole television programme was to turn on the television when the television company was broadcasting it, and sit through it to the end. By the 1970s, it was possible to record the programme on videotape and watch it after it had finished.

Today, there are many ways to watch programmes. Digital television allows many more channels to be broadcast. Television companies can use these to broadcast the same programme on different channels, with different start times. The viewer has more control with a set-top box that records programmes on a hard drive.

This allows the viewer to pause the action for a few minutes and return to it, to rewind and replay sections, knowing that the system will record the rest of the programme. These systems will also record favourite programs and allow the viewer to watch them on demand. (See also page 46.)

In the future, all television programmes will be available on demand. That means you will be able to choose which programme to watch and it will come to your television whenever you want it. These services will be delivered over the internet, when the speed of the internet is fast enough for high-speed, high-definition **streaming video**.

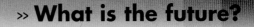

>> What is the future?

3D TV

Systems are now being developed that will show pictures in three dimensions (3D) instead of two. There are already 3D screens for mobile phones and laptop computers. They work by allowing the viewer to see two images, one for each eye. The screen divides the two images into vertical stripes. Each eye can only see one image. The brain combines the two to see a 3D image. However, there are many practical problems to overcome before 3D TV comes to our homes.

DVD and beyond

When you sit down to watch a DVD (digital versatile disc) or a video, or when you film something with a video camera or video phone, or record a television programme, you are using technology that records or plays moving images.

Moving pictures

Electronic moving picture, or video, systems, work by taking thousands of photographs of a scene in quick succession. This is normally 25 or 30 times per second, depending on the television system.

Each photograph is called a frame. To play a video back, the frames are displayed on screen in quick succession. Our eyes cannot react fast enough to see one image change to the next, so we see a moving picture.

Recording 25 or 30 pictures every second represents massive amount of information. It means that recording moving images is a far more difficult task than recording a still photograph.

A DVD and VHS video cassette player combined. It plays analogue tapes and digital optical discs.

Landmarks in time

1826 Nicéphore Niépce takes the first photograph using paper sensitive to light

1877 Eadweard Muybridge takes time-lapse photographs of moving objects

1889 Thomas Edison develops a cine camera and device to show films, called a kinetoscope

How video works

The job of a video camera is to take pictures or frames in quick succession and turn them into a signal. The camera has a lens that collects light from the scene and focuses it to make an image on a microchip called a charge-coupled device (CCD) (see page 31 to find out how CCDs work). Electronic circuits calculate the colour and brightness of each pixel in each frame and build a signal. The information for the next frame follows the last frame. To record the moving picture, this signal must be recorded on magnetic tape, on disc, or in the memory. Another device is needed to play the recording. It rebuilds the picture signal so that it can be sent to a television screen.

History of video recording

The first way of recording moving images was to take photographs in quick succession on a strip of film. This was done by a cine camera. Nobody uses cine cameras today, except some professional moviemakers. Electronic video camcorders became available in the 1970s. A camcorder is a combination of video camera and videotape recorder. Video recorders for recording television signals were introduced at the same time.

Cine cameras like this one recorded moving pictures onto a strip of plastic film. The film was processed to make the images appear, and played on a cine projector.

1951 The first videotape recorder is invented by the Ampex Corporation

1976 The first videotape cassette system, called Betamax, is launched

1980 Sony launches the first consumer camcorder

1994 Video recording onto CD is introduced

1997 The DVD video format is launched for the playback of pre-recorded movies

Where are we now?

There is a baffling range of video recording devices available at the moment. We have camcorders, videotape recorders, DVD recorders, and pre-recorded tapes and DVDs.

Camcorders

State-of-the-art camcorders record in the digital video (DV) format onto Mini-DV tapes. DV signals can be uploaded to a PC for editing. Tapeless camcorders are just appearing. They store video on a hard drive in the camera, or on mini DVDs. Camcorders have large LCD screens for composing shots and watching recorded video. They also offer many electronic special effects. Most camcorders also take digital photographs and record MPEG-4 video clips on a memory card. These can be stored and emailed.

Recording from television

We can record television programmes on videotape recorders, on DVD recorders, and on set-top boxes that record pictures on hard-disk drives.

The videotape recorders used in homes record analogue signals only. DVD recorders record signals onto recordable DVDs in digital form. The recordings are clearer than on videotape, and any part of a recording can be found instantly, without having to search backwards and forwards through a tape. Some DVD recorders also have a hard-disk drive that records signals, too. This allows video to be edited before being saved to the DVD.

DVR

A digital video recorder (DVR) records only to a hard drive. It can be used with television programmes only, and cannot play DVDs. It records digital television signals continuously, and can replay signals at the same time. This allows a viewer to pause or even rewind a programme while watching live, and then watch to the end of the programme as though it were still a live programme.

Pre-recorded DVDs

Films and other programmes are recorded on pre-recorded DVDs in a compressed (reduced in size) digital format called MPEG-2. MPEG stands for Motion Picture Experts Group. This organization agrees the standard formats for video and audio recording.

DVD piracy

Pirating means making a copy of a pre-recorded DVD and selling it to somebody else. DVDs can be copied on a standard PC with a DVD writer. Pirating is big business because the pirates sell copied DVDs cheaper than the originals. However, it is illegal, and the quality of the DVDs is often poor. The companies that make and distribute DVD movies are trying to find ways to stop pirating, but it is not easy. One possible solution is a disc that self-destructs after a certain number of viewings.

This device is an attachment for an Archos multimedia storage device. It records digital video in MPEG4 format.

Personal media players

Most devices that play video, such as DVD players, DVD recorders, videotape recorders, and camcorders, send out a signal to a television set. Portable DVD players play DVDs on a small LCD screen. However, a more complex device, the personal media player (PMP), has recently been developed.

This plays video recorded on a micro hard-disk drive and displays it on an LCD screen. It can also record video from other sources such as videotape players and DVD players. A PMP can also play MP3 files and store thousands of photographs.

The Archos Pocket Media Assistant (PMA) is a personal organiser and multimedia device that stores and plays video, photos and music files.

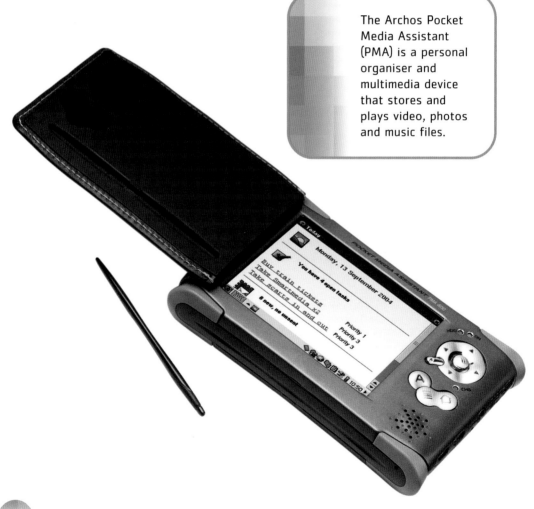

Home cinema

A home cinema system (or home theatre system) re-creates a cinema experience at home. A typical system is made up of a large widescreen television, such as a plasma or rear-projection screen, a set of surround-sound speakers, a DVD player, and a decoder for the surround sound.

In home cinema, sound is just as important as the picture. Surround sound plays different sounds from all around the viewer.

This makes viewers feel as though the on-screen action is happening around them. A state-of-the-art system has six speakers: five standard speakers plus a subwoofer, which plays the bass sounds, such as explosions.

Several channels of sounds are recorded on the DVD, one for each speaker. A device called a signal processor decodes the sound signal from the DVD player and sends it to the speakers. Some DVD players have this feature built in.

» What is the future?

The near future will see the next generation of **optical discs** for recording digital video. These will be important for recording high-definition television signals. There are two technologies in the pipeline, that have been developed by different companies. The first is high-definition DVD (HD-DVD), which can store 30 **gigabytes** (GB) of data (compared to 9 GB on a DVD). The second is called Blu-ray. This system uses a blue laser rather than a red laser to write and read the disc. A Blu-ray disc will hold a massive 50 GB of data. In the distant future, optical discs could hold up to 250 GB, enough for 120 hours of video.

The personal media player may become the heart of the home entertainment system, known as a media hub. It could double as a digital photograph album and digital music store.

Pong to PlayStation

A video game is an interactive game that is played on screen. A state-of-the-art video games console such as the Xbox 360 or PlayStation 3 is an amazing piece of electronic technology. It has many advantages over a PC for playing games: it costs less, is simple to operate, and can be easily connected to a television. Video games systems are not just for playing games. With the right software, they can also be used for education and training.

Video games in the past

The first video games appeared in amusement arcades in the early 1970s. These were too large and expensive for the home. The first successful home video game was introduced by the Atari company in 1975. It had just one built-in game, which was a simple bat-and-ball game called Pong. Two years later Atari introduced its Video Computer System, with plug-in memory cartridges for different games.

Playing the computer tennis game Pong on the Atari Videomaster games console. This was cutting edge in 1977!

Landmarks in time

1971 Computer Space is the first arcade video game

1972 Pong (a simple tennis game) becomes the first successful arcade video game

1972 The first video game console for homes is sold by the Magnavox company. It can play just one game – Pong

✕ Make the connection

How video games work

A video game system is actually a dedicated computer with powerful graphics. The main parts of a video games system are:

- A processor (like the processor in a PC) that runs the games programs. For example, the PlayStation 3 has a main processor called a cell running at 3.2 GHz.
- A graphics processing unit (GPU) that draws the graphics of a game, including special effects such as **texture mapping** and lighting. For example, the PlayStation 3 has a GPU that can draw about 75 million shaded **polygons** per second.
- Memory for storing the game program and game data.
- A user interface that takes data from a game controller and passes it to the processor.
- Game programs and data, which are stored on Blu-ray, DVD, CD or memory cartridges.
- Hard drives and memory cards to store game data, high scores and so on.

The system that started to make video games as popular as they are today was the Nintendo Entertainment System of 1985. This had graphics as good as those in arcade games. Since then, video games systems have become more and more powerful, with faster and more detailed graphics, and more complex games. The first handheld video games system that was successful was the Nintendo GameBoy, introduced in 1989.

1977 Atari launches its Video Computer System, with games on cartridges

1978 The classic game Space Invaders appears on arcade consoles

1985 The Nintendo Entertainment system (NES) is launched and becomes a runaway success

1994 Sony launches its PlayStation in Japan, with games stored on CD-ROM

2000 Microsoft launches the most powerful games system so far, the Xbox

Where are we now?

The three consoles at the cutting edge of video games are the Sony PlayStation, the Microsoft Xbox and the Nintendo GameCube. These are all very powerful machines that can produce staggering three-dimensional moving graphics. As well as the consoles themselves, there are extra pieces of equipment, such as steering wheels, dance mats, and even drum kits, that can be plugged into the consoles for specific games. By using a network, computer-game players can play head-to-head over the internet. All these systems display images on a television screen.

The PlayStation Portable (PSP) is a state-of-the-art handheld game. It is a miniature system with all the graphics power of the large PlayStation. The fastest-growing games platform at the moment is actually the mobile telephone! There are thousands of games in the Java programming language that can be downloaded onto mobile phones. Games are becoming more and more sophisticated, with bestselling titles from games consoles being released in mobile-telephone format.

However, video games systems are just useless boxes of electronics without the games programs to go with them. The latest action games have stunning graphics that squeeze the most power from the systems.

The Sony PSP is the most powerful handheld games console so far, with the processing and graphics power of a PlayStation 2.

Video game problems

Many video games contain violent content, where players have to kill people, animals, or aliens. Children often play these games, which are made for adults. Many people are concerned that children could repeat what they see in games in real life.

So far there is no proof that this happens. People also worry that children spend too much time playing video games. They think they should spend some of this time playing active games outside, which help them to keep fit.

>> What is the future?

Most video games consoles have a life of about five years before they are replaced by a new, faster, more powerful console. This means that the current generation of video games consoles will soon be obsolete. Each new generation of consoles has more processing power and can produce more detailed graphics. This allows game programmers to create more realistic gaming experiences. These new games will require a huge amount of data to describe the worlds they are set in. The consoles will therefore need the next generation of **optical drives**, such as Blu-ray and HD-DVD (see page 49). In the future, games consoles will connect to the internet to allow games to be downloaded, and will also allow us to send emails and surf web pages.

Glossary

amplitude size or strength of a wave or signal

analogue relating to a device or signal in which information is represented by something that can have many different values or levels

application software that allows a computer to perform different jobs, such as word processing or games

bandwidth speed at which information can travel along a communications link, such as a cable or optical fibre

binary number number made up of just zeros and ones

Bluetooth a type of wireless communication used for short-range links, such as headsets with telephones

broadband a communications link with a high bandwidth

broadcasting sending sounds or television pictures from one place to many radio or television receivers

cathode ray tube (CRT) vase-shaped glass tube used to produce the picture in a television

CCD (charge-coupled device) integrated circuit that detects light in a digital camera or camcorder

CD (compact disc) plastic disc for recording digital information. The information is recorded as pits in the plastic and read by a laser.

CD-R short for compact disc recordable. You can record information onto a CD-R just once.

CD-RW short for compact disc re-writable. You can overwrite information on a CD-RW many times.

component a device that is part of an electric circuit, such as a resistor, transistor or LED

compression way of reducing the space needed to store digital information

download to move information (such as MP3 files or photographs) from a digital device to the device you are working on

digital relating to a device or signal in which information is represented by the binary numbers zero and one

DVD (digital versatile disc) plastic disc for recording digital information. The information is recorded as pits in the plastic and read by a laser.

electronic relating to a device that works by electricity and processes signals in some way

format a standard way of recording information

gigabyte (GB) unit of memory in a digital device. It is equal to 1024 megabytes.

GPS (Global Positioning System) a navigation system. A GPS receiver uses signals from GPS satellites to work out where it is to within a few metres.

graphics any image drawn by a computer

hard-disk drive (HD) device for storing digital information

JPEG digital photograph format used to store photographs. JPEG stands for Joint Photographic Experts Group.

LCD liquid crystal display: a screen in which each pixel is a tiny crystal that blocks light or lets light through

megabyte (MB) unit of memory in a digital device. It is equal to 1,048,576 bytes

megapixel a measure of the number of pixels on a CCD or in a digital image. Equal to one million pixels.

memory stick a digital memory device that normally plugs into a computer

microchip small piece of silicon with many electronic components built into it

microprocessor (or processor) integrated circuit that is the 'brains' of a digital device such as a camera, mobile telephone, or computer

MP3 digital music format used to store music in digital music players

network many computers or other devices connected together so that they can swap information

optical disc a disc such as a CD or DVD from which data is read by a laser

optical drive any digital storage device that uses discs written and read by a laser, such as CDs and DVDs

optical fibre a thin strand of glass along which signals are transmitted as flashes of light

palm-top computer a computer small enough to fit in one hand

pixel picture element: one of the tiny dots that makes up an image on an LCD screen or digital picture

plasma screen television screen in which each pixel is a tiny glass bulb full of a substance called plasma.

personal media player (PMP) digital device that can store, play and display digital music, photographs and video

polygon closed shape with three or more straight sides.

receiver device that detects radio waves and turns them back into sounds or television pictures

signal a changing electric current or radio waves that represents information, such as sound, images, or data

smartphone device that is a mobile telephone and an electronic diary in one

streaming video video that is watched as it downloads into a digital device such as a mobile telephone

surround sound a sound system that plays sound from speakers all around a listener

telecommunication sending and receiving information such as sound, text and video by electricity, radio and light

texture mapping in computer graphics, painting a pattern onto the surface of a computer model of an object

transistor electronic component that works as an electronic switch to control the strength of an electric current

transmitter device that sends out radio waves

upload to move information (such as MP3 files or photographs) from the digital device you are working on to another digital device

valve short for thermionic valve. A device used as an electronic switch before the invention of the transistor.

Wi-Fi (wireless fidelity) network that uses radio waves instead of cables to link computers and other digital devices together

WMA (Windows Media Audio) digital music format used to store music in digital music players

Further resources

Tabletop Scientist: Sound, Steve Parker, Heinemann Library, 2005

Everyday Science: Electricity, Steve Parker, Heinemann LIbrary, 2004

Index